MY BIGFOOT LIFE
Adventures in search of Britain's biggest mystery

Daniel Barnett

Typeset and edited by Guinevere Palmer,
Layout by Muhammed and Ibrahim for CFZ Communications
Using Microsoft Word 2000, Microsoft Publisher 2000, Adobe Photoshop CS.

First published in Great Britain by CFZ Press

**CFZ Press
Myrtle Cottage
Woolsery
Bideford
North Devon
EX39 5QR**

© CFZ MMXXIV

All rights reserved. Without limiting the rights under copyright reserved above, no part of this publication may be reproduced, stored in or introduced into a retrieval system, or transmitted, in any form of by any means (electronic, mechanical, photocopying, recording or otherwise), without the prior written permission of both the copyright owners and the publishers of this book.

ISBN: 978-1-909488-71-7

DEDICATION

I would love to dedicate my book to my amazing Grandparents who have helped me to achieve what I have accomplished in the Bigfoot world.

My amazing Gramps sparked my interest in the mystical universe around us. My Gramps is the inspiration to my research, and I will never forget the first day we sat on the sofa, watching expedition television shows together. That day started a whole new world of discoveries; he is a brilliant man. My journey would not have been possible without him by my side.

Now to my amazing Nan, she is an incredible woman who has explored with me on every expedition; every time I need her, she is there. She joins me on my expeditions, encourages me, invests her time in me and always believes I can achieve my goals. This book wouldn't be here if it wasn't for her, and she needs to be recognised for her enormous heart and her endless love for her grandchildren. Thank you for trudging round the forest with me.

She secretly loves it.

Thank you for everything you have both done towards my research within the wonderful world of Mythical creatures.

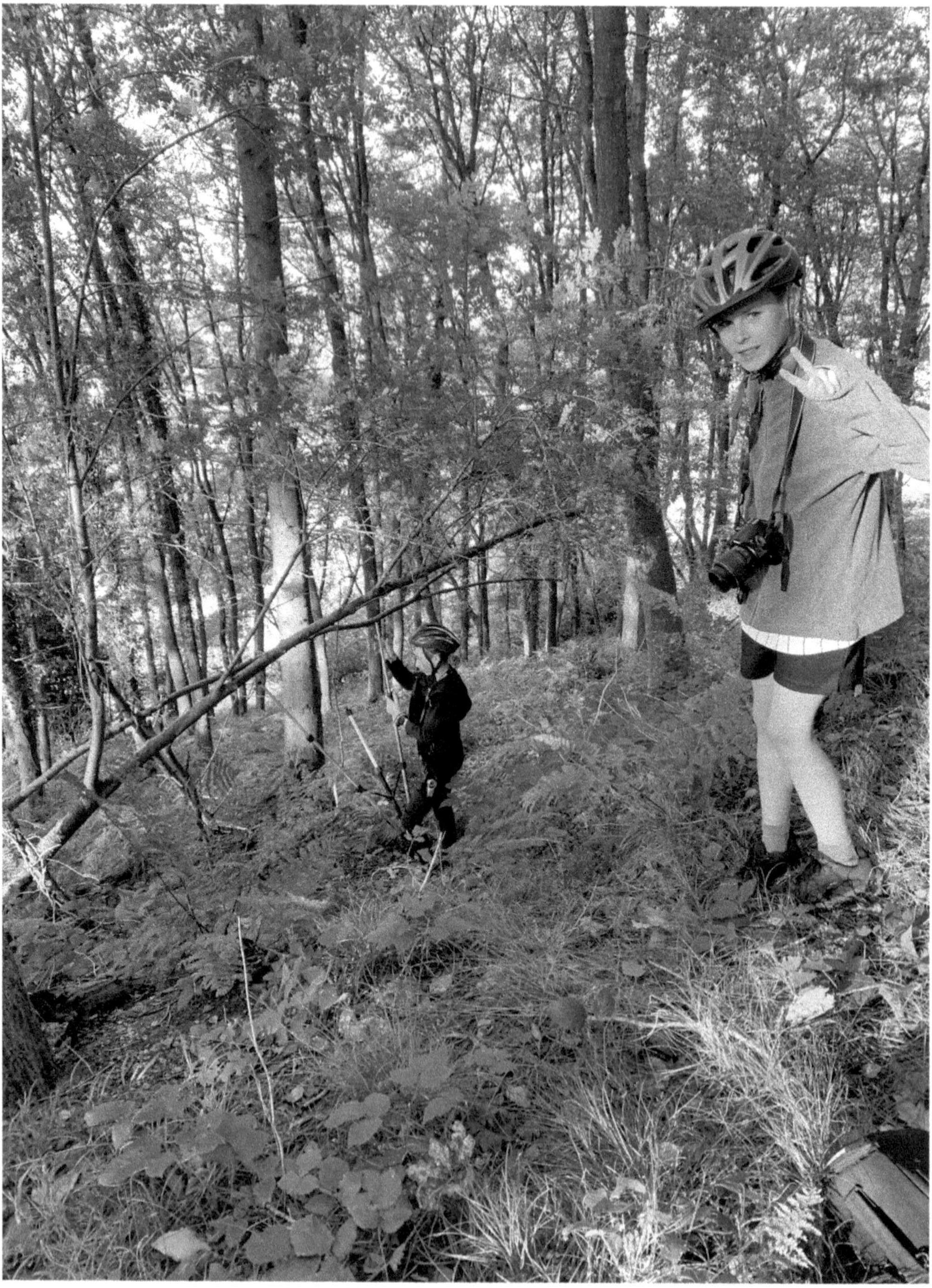

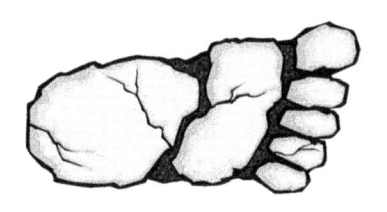

FOREWORD

When I stumbled across Daniel Lee on a Bigfoot Facebook page, I knew that I had found someone special. This young man has a sense of curiosity and wonder that is sorely missing from our world today. When he has a goal, he chases it down and turns it into reality. When he has a theory, he squeezes every bit of truth out of it, yet he remains open minded. He is not afraid to be wrong, but he is hopeful about his potential to be right. Daniel doesn't jump the gun when he comes across potential evidence. Does he get excited? Yes! But in the middle of his excitement is a real desire to know the truth.

He is not afraid of disappointment, because every debunked piece of evidence brings him knowledge. So, he brings his findings to others, and he asks them to help him make sense of it. I've had long conversations with Daniel about why I don't think something is evidence of Bigfoot. With most young people, I might want to let them down gently, but Daniel craves realness. So, I am real with him, as I would be with any colleague two or three or four times his age. Daniel knows that I still have my doubts about Bigfoot being in Great Britain. Before I met him, I had never really given it a second thought. Then again, I call myself a hopeful sceptic on the topic as a whole. But here is what I know: Daniel Lee is honing his skills and is becoming a stellar researcher. I am proud to know him and his dad, Craig, and to cheer them on from across the pond. I can't wait to read his book and to see what research he's involved in down the road. Because if anyone has the grit and determination to make known the unknown, it is Daniel Lee.

Sometimes as we get older, we forget about what drove us in our youth, like our passion for the unknown. In "My Bigfoot Life" Daniel reminds us to embrace our passion for the Unknown. With each chapter Daniel documents his journey and adventures into the subject of Bigfoot. As the reader, he reminds us we can learn at any age and reach out to others to help map our journey to the discovery of the unknown. So, no matter if you are new to the subject of Bigfoot or someone who has been researching for years, sit back with your favourite trail mix and enjoy Daniel's journey into the subject.

Mike Anne & Amy Bue
December 2023

CONTENTS

DEDICATION 3
FOREWORD 5
Introduction 7
Chapter 1 – How it began 9
Chapter 2 – Expedition Deer 11
EXPEDITION DIARIES 13
Chapter 3 – Mythical Legends Podcast 23
Chapter 4 – Bigfoot Researching 25
Chapter 5 – Shocking revelations 29
Chapter 6 – Current Investigations 31
Chapter 7 -Little Feet 37
Chapter 8 – New Researchers 41
Chapter 9 – British Bigfoot 43
Mythical Legends/Little Feet in the news 47
Chapter 10 – What to do if I come across evidence? 59
Chapter 11- The Future 61

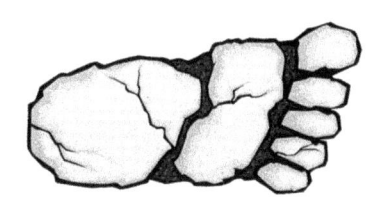

INTRODUCTION

My name is Daniel Barnett. I am 14 years old, and I am the UKs youngest Cryptozoologist and Bigfoot researcher.

I am going to take you through my journey searching for this elusive beast. Our adventure began the summer of 2022 and I have to say, I am astounded by the stories leading to the Bigfoot mystery. Embarking on this expedition has taken us 18 months so far. It started with my family helping out and it grew to having my own team. My family members consisted of my parents, Craig and Gemma Barnett, my sister, Emily Barnett, my grandparents, Jill and Adrian Roberts and my cousin William Roberts.

Bigfoot legends go back beyond recorded time, it's incredible and I can't wait to share my experiences and my journey about the researcher I have become.

So, sit back and open your mind, because we are in for a long ride of puzzling, mystical events that are, I assure you, all 100% real!

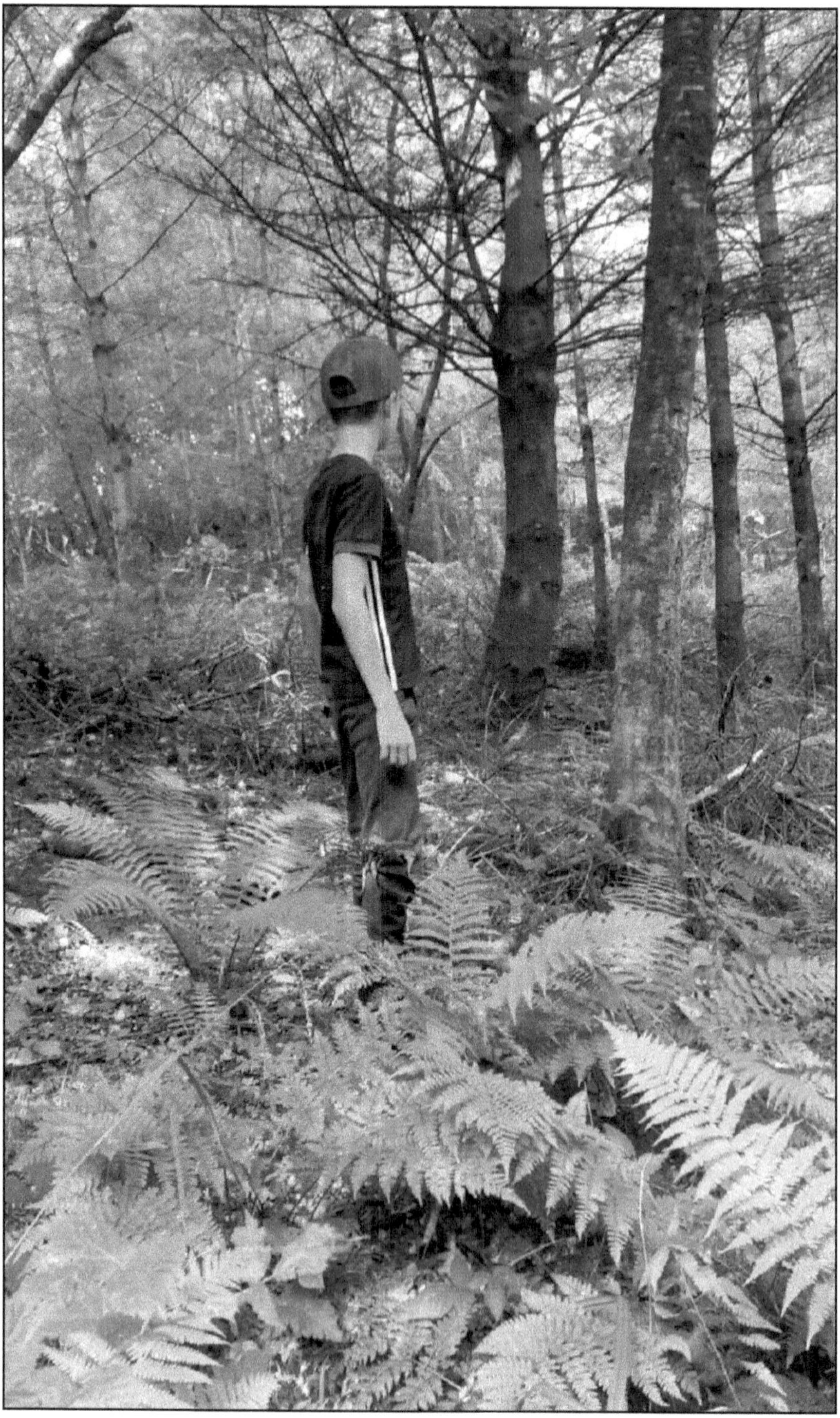

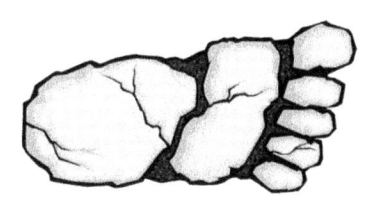

Chapter One
How it began

Back in 2021, I was sat with my Grandad and cousin, and we were watching a great TV program called Expedition Bigfoot[1]. We watched the first episode and when we went away to talk about it, we couldn't stop talking about it for hours. When we saw that the next episode would be on the following week, we were so excited. During that week I was looking into what Bigfoot was and without realising, I had opened up a whole new world. Ahead of me was a big, huge town full of researching, talking, reading and watching.

Before I knew it, it was time for my cousin and I to make our way to my grandad's house to watch the next episode. It got me more and more fired up to research and at this point I had no intentions of going out into the forest and becoming a cryptozoologist. It's crazy to look back and see, that I didn't even dream of doing what I do now. After that episode I decided to do a bit of researching online about what Bigfoot is. I started to scroll through all these sightings about Bigfoot and, funnily enough, I found out how big his feet were (the answer is around 18 inches) and it shocked me!

It was two weeks later, and I decided to go back and watch the first two seasons of Expedition Bigfoot, so I was very caught up on what they were talking about. At the end of season three, we were all gutted!!! So, we were talking about the last episode and how it was very unfortunate that they didn't see what was coming in and out of those nests. I was speaking to my grandad, and we were discussing about other Bigfoot programs. I mentioned to him about Bigfoot in the UK, and he said 'no, that's impossible', and from then on, I thought the same.

---
[1] https://www.imdb.com/title/tt11274284/
[2] https://www.facebook.com/TheJamesBoboFay

It was then three weeks later, and I was talking to my nan about all the Bigfoot facts and the interesting sightings. It was a day later when I thought to myself, has anyone actually seen Bigfoot here in the UK? and when I searched for it on Google, the answer was yes, it was a video of Bobo[2] and his team, coming down to meet a man called Neil Young, because he had a Bigfoot sighting, and my jaw just dropped. I went straight round to tell my Nan about it and she said, 'that's incredible!'. It was a week later when I said to her, I need to go out in the forest and start gaining experience and start to learn more about what all of this means. So, I talked with my cousin, and we eventually found a place to go and gain the skills we needed. That was when my first YouTube program was born: "Expedition Deer[3]".

---

[3] https://www.youtube.com/@MythicalLegends282

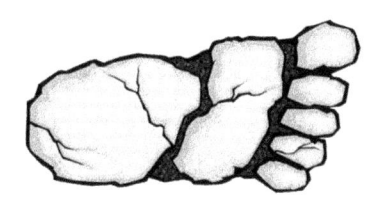

Chapter Two
Expedition Deer

As soon as we could, both me and my cousin got some hiking gear and along with my Nan and Grandad, we jumped in the car and went to the nearest forest we could, and that's where it started. At this point we were looking for deer, because we didn't have enough information to go out looking for Bigfoot yet. We were out using our rope gaining skills and we were having a good time of it.

When the time came to an end, I turned to my cousin and said, 'we can't stop this we need to continue, we need to research as much as we can,' so that's absolutely what we did. We set up our small team within our family and we were out every two months or so, looking for deer, gaining as much experience as possible and learning from all sorts of mistakes (don't forget we were a little younger back then, so we were also messing about) and we were having the time of our lives.

On a cold winter's morning, Gramps and I were up on a ridge, where we caught something running up on another ridge in front of us. I said it was probably a person, but then as gramps looked closer, we confirmed it was a deer shooting by, it was a very rewarding moment to feel that mine and my cousins work had finally paid off. I stood there in pride and joy that we had gained enough research to start expanding. So, we decided we were going to do one more expedition but include everything we were ever going to need. So, when we went home, we bought plaster of paris and basic camping equipment and we went for our last expedition. We were out at 6am looking and within five minutes of us being there, we are suddenly stood in front of a roe deer. We are all just astounded, it was incredible, it was an incredible experience. All the work we had done brought us to this point. The deer even barked five times in front of us and you don't see that every day.

So, we came back home after we had that amazing encounter and we decided to end Expedition Deer. We have now been out many times, learnt a lot about our environment, the importance of being quiet and observant and how to use plaster of Paris. It does get to the point where you say 'how many times can we go out and do the same thing'. But from there, we came up with an even better idea!

EXPEDITION DIARIES

Thursday, 17th August 2023

Windown Forest - Mixture of forest surrounded by open fields.

Lots of game trails. Lots of evidence of broken twigs, leaves etc have been broken and crushed.

There were several water sources close by.

We heard lots of birds, mainly crows, magpies, plus unidentified calls.

We heard calls that we thought were likely deer calls, coming from different directions in the forest.

We found many single burrows, then came across a group of around 8 or 9 very large burrows which seemed to be linked to a possible underground chamber. We thought these were most likely made by badgers.

Throughout the day we came across many deer prints.

Whilst following a game trail we discovered two large indentations which looked like footprints. These were approximately 18 inches long. We took pictures of them and will ask others for opinions on what they could be.

We also came across what looked like a large paw print, possibly cat like. We took pictures of it and will ask others for opinions on what they could be.

Sunday 20th August 2023

Returned to Windown forest and retraced our steps.

We examined the indentation that we found on Thursday and took a plaster cast of it.

We found what looked like more footprints/indentations near the badger sets. We also found a small amount of white strands which we could not identify as either hair or fungi. We took a sample to look at under a microscope at home.

We did a quick check of the trail camera, which didn't show anything significant, so left it for another couple of days plus put a second one near the badger set.

We then removed the plaster cast and took soil samples for DNA testing.

Tuesday 22nd August 2023

Back into the forest today we explored an area we had not been to before.

We went as far as we could off the main track but did not find anything other than deer trails.

We returned to the area we had been in before to take down and check the trail cameras. Numerous pictures had been taken, some appear to show animal eye shine, most likely belonging to badgers. One photo shows a larger grey shape which we think is a deer.

Away from the first imprints we found, we came across what appeared to be 5 or 6 large indentations, very similar to the first ones we found. We took photos, video and samples for eDNA.

We also found more white strands, possibly hair, that we collected samples of.

We also found some scat, which we could not identify. It was almost black in colour, 3 round shapes, 1 approximately the size of a tennis ball, the other 2 slightly smaller. We took samples which include a hair from the middle of the largest one.

We left some apples placed on sticks plus we spread some peanut butter on some of the tree trunks.

Monday 4th September 2023

We returned to Windown forest; this time William accompanied us. We set up a tent and camping chairs and we planned being able to stay the whole day.

We found several more imprints which may or may not have been footprints.

We explored an area where there were steep slopes and discovered a tree structure and an arrangement of sticks laid on the ground with a small rock and stone balanced either end. We took photos of these.

William took several more photos of the area which Daniel later posted on Facebook groups. We had a message from a group member asking what the black shadow like shape was amongst the trees.

We had not noticed these when we were there. The shadowy figure resembled a large cat or bear.

The apples we had left last time were still there, but all the peanut butter was gone.

Saturday 23rd September 2023

We held our very first Bigfoot event!

The idea was to try to encourage local children to find out some interesting and fun facts about Bigfoot, also to encourage them to look at things outdoors, whether in the forest or parks, even their own gardens.

We had various activities including colouring competitions, crosswords, word searches etc.

We also had an exhibition table where we had lots of pictures of Bigfoot, tree structures, maps of sightings from around the world. We also included equipment to take to the forest, such as trail cameras, camouflage sheets, binoculars, tape and eDNA kit.

We gave the children notepads to take away to do their own mini searches.

The event was held at a local café, The Purple Spoon café at Highbridge, Somerset.

The event was a real success.

Sunday 1st October 2023

Following the results of the eDNA, a newspaper took up our story.

A photographer asked us if he could take photos of the forest where the soil samples were taken.

These photos were then printed along with a story in the Daily Star, out local Western Daily Press, plus several other newspapers.

Thursday 5th October 2023

We returned to Windown forest to see if we could identify exactly where the dark shape in one of our photos was, we went to the exact location the photo was taken expecting to see a tree stump or bushy shrub of some kind that could explain the shape in the photo.

Despite being there for almost 2 hours taking lots more photos and looking from every angle, we could not find anything to resemble that shape.

We realised that the steep slope meant that what we originally thought was a shape on the ground, could not be possible. The position from the photo meant that, due to the slope, the shape would have been high up in a tree.

We could not come up with any explanation for this.

Saturday 14th October 2023

We visited a different forest tomorrow. This is known locally as Seven Sisters.

It is a large area of several forests and connecting open grassland.

We went off the main pathways and came to a small clearing that had several possible tree structures around it.

We hound a couple of indentations that could have been large footprints. We took soil and leaf samples which included several hairs.

Sunday 22nd October 2023

We finally received the report from the lab in Portugal showing that the eDNA samples contained, among many other things, traces of Old World Monkey and Great Ape.

We were completely shocked at this result!

Friday 17th November 2023

Back to windown forest again, this time with Darren and Angela from South Coast Ghosts, who were attending our 2nd event the next day.

They were interested to see where we had explored, and we wanted to hear their opinions.

We took them to all of the places we had previously been, including the indentation/footprint where we took the first soil samples.

We then went to a denser part of the forest where Angela discovered a pile of bones. We photographed them and later were told they were the bones of a deer. We couldn't explain why they were there.

Darren and Angela both felt a certain vibe from the forest in general.

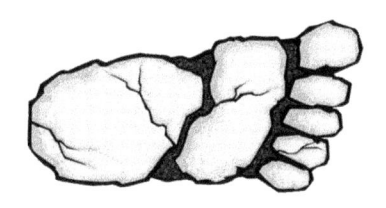

Chapter Three
Mythical Legends Podcast

After we finished Expedition Deer, we decided to continue our research and I said to my cousin and my nan, it was time for us to start researching Bigfoot.

We decided that we needed to get to know some of the researchers and scientists in the field, so we started Mythical Legends podcast[4].

We have interviewed some amazing people and learnt lots. The Bigfoot Community is huge and by knowing who to go to for advice is one of the key things you need when going into researching a cryptid.

We have spoken to some amazing researchers and scientists such as Amy Bue, a Bigfoot researcher and Co-Founder of Project Zoobook[5] and Shane Corson from the Olympic Project. I also interviewed some witnesses who had class A sightings of this creature.

We have interviewed Dr Jeffrey Meldrum[6], a professor of anthropology and anatomy. He is, in my opinion, the leading scientist in the Bigfoot mystery.

I have also interviewed Bill Munns; he is one of the only people to analyse the Patterson/Gimlin film and to say that this sighting wasn't a man in a suit. The Patterson Gimlin Film was filmed in 1967, when Roger Patterson and Bob Gimlin filmed a creature, that was later nicknamed Patty, walking across a sand bar.

There have been many attempts to debunk this film, but no one to this day has disproved it. The question remains whether this film is real evidence of

---

[4] https://open.spotify.com/show/189JGCISFFIXQy8IUOmA5G?si=4b6bf774dd8440a0
[5] https://podcasts.apple.com/ca/podcast/amy-bue-bigfoot-and-project-zoobook/id1518903701?i=1000514649350
[6] https://www.isu.edu/biology/people/faculty---professors/jeffrey-meldrum/

what is actually out there. This film was filmed in Northern California in Bluff Creek. There have been certain frames pointed out by Dr Jeffrey Meldrum and Bill Munns that show that it could not have been done with a man in a suit. Technology advances that show for example, the way the foot moves and the way the back has primate features, and this is very interesting to Dr Meldrum and Bill Munns.

Some of the frames in the film are particularly intriguing as they show the creature doing things that normal Humans can't do, such as the hand moving from side to side and the hair on it's hips seeming to move and change colour. The only one of the men who took it left alive today - Bob Gimlin - is the most respected man in the Bigfoot community. Without him we would be unable to carry out the work we are doing today. I give my respect to Bob and I thank him for this.

Talking to Bill was an incredible experience and Dr Jeffrey Meldrum was amazing. Maybe, just maybe, there is really something out there that we haven't discovered yet.

Their vast range of knowledge was extremely interesting with lots of new topics coming to light.

I also found that sometimes people have been very scared by their experiences and have not been able to tell a soul until maybe 35 years later. So, whatever they saw it must have scared them very much.

They have inspired us to change some of our ways of exploring the vast wilderness, such as the silent sitting treatment.

By the way, if you wanted to listen to my podcast, it's on Spotify, Apple podcasts, Amazon and lots more!

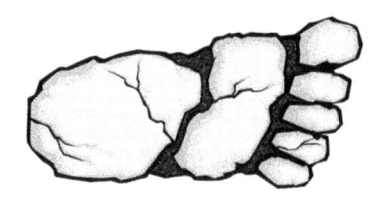

Chapter Four
Bigfoot Researching

Once we had enough information from the Podcast and we had all the knowledge of being out in the forest, we decided it was time to arrange which forest we were heading to. We had picked a location on the Quantock hills[7], and we had picked an area where there was a nearby water source. We chose an area where humans don't tend to go surrounded by thicket. Once we picked a location, we prepared all the equipment we might need, and we set off to our secret location.

I have a few opinions that I have about the cloaking theory and why Bigfoot makes tree structures.

When people bring up Bigfoot a lot of people refer to something called "Cloaking" suggesting that like the characters in a book like Harry Potter the creatures can appear or disappear at will. I believe that these people are basically right but that they are not using the right term of phase. I use the word Camouflaging, because I am know that things that live among us today that do exactly that. The animals I am referring to are Chameleons which have specialized cells in their skin called chromatophores that contain pigments which can change colour according to the surroundings. These chromatophores are controlled by the chameleon's nervous system, allowing them to adjust their appearance rapidly. If we relate that back to bigfoot it wouldn't be out of the bounds of possibility to extrapolate something like this. It would also be logical if these animals are using items from their environment to cover themselves and using the undergrowth in which to hide. Either, or a mixture of *both*, would be able to explain why people say the go invisible.

---
[7] https://www.quantockhills.com/

When we arrived at the forest, our algorithm was right; there were no people there at all. So, we set off into the forest and immediately we start hearing sounds, some we could identify and some we could not. Some noises sounded like deer, (we weren't going there just to search for Bigfoot, we went there to see what we could find) we went to see what it could be, and we couldn't find anything, so we moved off the human path and moved into the forest itself.

When we moved in, we saw this massive ridge, so we decided to climb the ridge and when we got to the top, I started looking around for deer. My partner who was with me, called me over and in front of us was a huge 18-inch print. We both stood there in shock and no words came out of our mouths for a long time. We were excited and were trying to anticipate what it could be. When we eventually did speak, we got our cameras out and took lots of pictures. We made a cast with plaster of paris and we took some eDNA from this print.

eDNA is an amazing piece of equipment if you know how to use it correctly. Taking eDNA samples can be very complicated but I'm going to take you through the steps so you can also use this on your research. Its ideal that you take this kit out with you every time you go out in search for something. People think you need lots of money to have all the fancy tools but truly, that is not the case.

So let's start with your average kit, you should have:

- At least 2 pairs of Latex gloves. You never know how many pieces you will want to collect and you also want to try and keep your DNA off of the sample you're collecting.
- Either test tubes or sealable bags to put your sample in. Two bigger sample bags in case you want to collect a bigger sample.
- Two (or more) different tweezers so you don't mix up your eDNA.
- A pen and labels for you to label your eDNA and where you found it.

Then once you have your kit you are prepared to head out into the forest.

When you come across something that you think may have DNA on and you want to collect a soil sample, then this is what you do.

- Put your bags or test tubes to the side to stop them falling into what you are collecting and, if possible, open up the top of the bag or tube so its easy to put the sample in if you're on your own.
- Put on your latex gloves.
- Grab your tweezers by the top and not the end you are picking the sample up with because you don't want your DNA on the sample.
- Then you get your partner to open up the bag and test tube by the top, again, don't let your partner put there hands in, because you don't want to contaminate your sample that way.
- When you have completed that step you can then begin to pick up your soil sample and put it in the bag or test tube. Once your soil sample is in the bag/test tube you then need to shut the bag/test tube and put it in another bag. Then label it with where you found it, the date you collected it and any other useful information. That way, when you send it off for testing, you can let the scientists know.
- Once you have finished that stage you are safe to take your gloves off and dispose of them (responsibly!) Put your eDNA somewhere safe so you can continue your investigation.
- As soon as you get back from where you collected it, you need to put it in your fridge or in a cool environment, until you are ready to send it off.

GLOBAL BIOSCIENCES CENTER

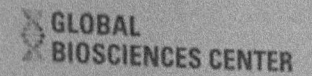

SGS

Samples & Results

On the 30th of August 2023 the SGS Global Biosciences Center received 5 soil samples for Chordata taxonomic identification through E-DNA metabarcoding analysis. The following samples where provided.

Sample	Description	Measurements (cm) (L x W x D)	Expedition day
BFT01	Footprint - Game trail	33 x 15 x 4	E1
BFT02	Footprints	31 x 24 x 3	E1
BFT03	Footprint cast vegetation		E2
BFT04	Drop down from sets footprints	31 x 19 x 4	E3
BFT05	Bipedal movement on path	18 x 20 x 4	E3

The analysis of all samples was performed successfully and although initial sample volume was low, all samples yielded enough quality DNA that allowed the identification of the taxa listed below through eDNA metabarcoding. When species identification could not be made with high confidence, a higher taxonomic level - genus or family level - was listed. Failure to identify a species with high confidence may occur if the DNA molecules are partially degraded not allowing for a correct identification and/or if the reference databases do not have the enough genetic information to discriminate the organisms at a species level. This may be the case of the taxa highlighted in the table below (*), listed at the genus or family level.

Taxa	Classification Level	Common Name	BFT01	BFT02	BFT03	BFT04	BFT05
Columba palumbus	species	Common Wood Pigeon		detected	detected	detected	
Gallus gallus	species	Red junglefowl (chicken)			detected		
Phasianus colchicus	species	Common Pheasant		detected	detected	detected	
Capreolus capreolus	species	Roe Deer				detected	detected
Sus scrofa	species	Wild Boar (or domestic pig)			detected		
Canis	genus*	Wolf, dog	detected		detected	detected	
Canis lupus	species	Wolf (or domestic dog)	detected	detected	detected	detected	detected
Meles meles	species	European Badger				detected	detected
Cercopithecidae	family*	Old World Monkeys	detected	detected			
Hominidae	family*	Great Apes			detected		detected
Homo sapiens	species	Humans	detected	detected	detected	detected	detected
Sciuridae	family*	Squirrels					detected
Neosciurus carolinensis	species	Eastern Gray Squirrel			detected	detected	detected

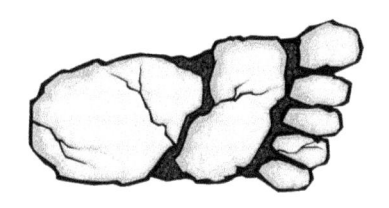

Chapter Five
Shocking Revelations

After finding weird prints, we took more eDNA from indentations below these huge prints. When we returned to base, we decided to start asking around, to see if anyone would accept our DNA samples. There was no one to start with, until I had a phone call to say that a company had accepted our samples. We shipped them off and they said it would take 6 to 8 weeks for the results to come back. So throughout that time, we were investigating the forest and finding all sorts of weird things. We found lots of tree structures, and they were huge, we found deer carcasses and a massive pile of unknown scat.

We had updates constantly from the company analysing the eDNA until they said a few more weeks and it will be done. Well, those few weeks went by so fast, and it was time for those results. My dad received a call, and the results they told him shocked him. He then went straight on the phone to me, and he said they have found possible traces of old world monkey[8] DNA. I stood frozen for about 3 minutes because I could not believe what we were hearing. When the final report came through, there were even more shocking revelations. On the report it had old world monkey and great ape[9]. The weird thing about this, is the company could not narrow down what species the monkey was.

Old World monkeys are primates in the family Cercopithecidae. Twenty-four genera and 138 species are recognized, making it the largest primate family. Old world monkey genera include Baboons (genus Papio), Red Colobus (genus Piliocolobus) and Macaques (genus Macaca).

---

[8] https://en.wikipedia.org/wiki/Old_World_monkey
[9] https://en.wikipedia.org/wiki/Hominidae

Phylogenetically, they are more closely related to apes than new world monkeys. Old World Monkeys and apes diverging from a common ancestor between 25 million and 30 million years ago.

It was weird as hell!

Many people believe that Bigfoot creates the numerous tree structures people find in the woods. There is no direct evidence that this this the case, but I have a theory for why it might be.

We always ask ourselves when we go out in search for this elusive beast: Where are Bigfoot Bones? My answer to that is that there are other animals alive today that seem to copy our movements or at least behave in similar ways to us - for example the three species of orang utan. The creatures use some of the same body gestures and these creatures even use tools to help them eat. To refer back to Bigfoot, if you think for a moment that they could bury their dead, and if so, and if they are responsible for the tree structures, could these be grave markers? Bigfoot might have learned that skill from us or might have developed it before we were around. Primordial man might even have learned it from them! The question remains has anyone ever dug under a tree structure?

Don't get me wrong, tree structures could also be used for communication and marking territory, or they could have nothing at all to do with Bigfoot, but it's a theory to start thinking about.

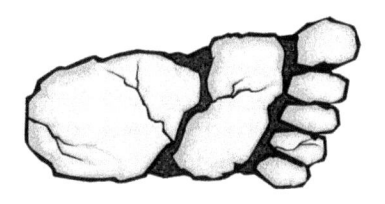

Chapter Six
Current Investigations

I am currently working on a new project called 'Little Feet'. We are very interested in getting young children involved in Bigfoot researching. So, our main aim is going to be getting these young children out into the forest, looking for signs of Bigfoot, but also to be getting outside with their family. We are putting on events every couple of months or so, with lots of researchers coming down to give talks to the adults and children. We also have a massive kid's area, for them to learn a lot about what Bigfoot actually is and to start to get to know what we do.

As an expedition team and as a group called Mythical Legends, we are very proud of what we have done and the amount of learning we have completed is incredible. Thank you to those who have helped out with that, we really appreciate it. So, from this page, go on and find out about all the interesting facts we have learnt and learn more about Bigfoot.

Other than that, keep searching and stay Mythical.

This is some unknown scat we found next to a possible tree structure

My Bigfoot Life

This is a shadow that appeared on our photo

This is another possible structure that we came across

This is a tree structure that we found, it was 15 feet high

This is an Indentation we found further up the trail from the 18 inch one

These are toe impressions we found and something about it we found weird

My Bigfoot Life

This is the cast from the 18-inch footprint

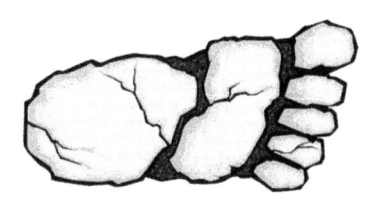

Chapter Seven
Little Feet

Without many people knowing here in the UK, we have a Cryptid Convention running and I would love to discuss it.

I have put an organisation together called 'Little Feet' and as an organisation we work together to bring similar people together in one room.

Our secondary aim is to get children out into the forest looking for these mysterious creatures and to bring more people into the Cryptid world.

We had our first convention on Saturday 18th November, and it was a smashing success. We had many talks, and it was a very enjoyable day. We will never forget the first ever event.

We are encouraging more and more people to attend this event yearly, to find out what we are all about and find out what is actually out there. I say "There is no harm in finding out what is lurking out there." When I say that, some people may think I am mad, but it really means something when you think about it.

When you are on Facebook and you are looking for something to do, find out when our nearest event is going to be, because you never know, we could have something exciting going on just around the corner.

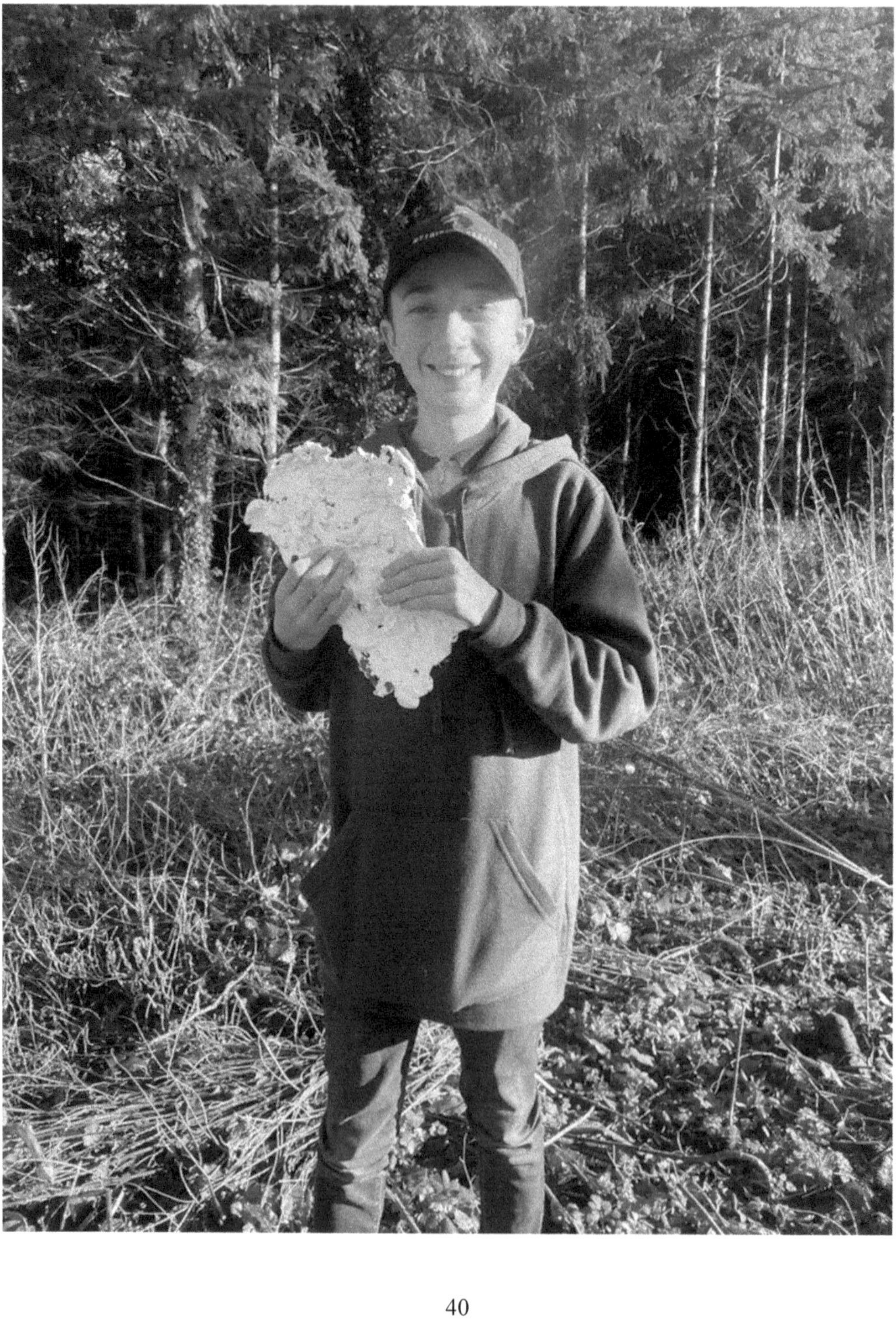

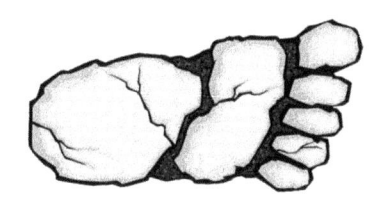

Chapter Eight
New Researchers

If you are sat at home reading this book, or you are sat watching some of the work that we have done, and you decide you want to go out looking for this mysterious beast, then keep reading because I am going to advise you on what to do.

Before you jump ahead and go straight into the forest (which might sound like a great idea at first) there are a few things you might want to take into account.

My number one top rule is always go out prepared. The reason why I say that is because, let's say, you come in front of this creature, what are you going to do?

There are many things that could go very wrong if you are in the presence of this creature.

My suggestion would be, go and do some research online and find out what encounters people have had, so when you eventually go out into the forest, you are not struck with unfamiliar situations.

My number two rule is always know where you are going.

If you turn up to a forest and you have no idea what you could stumble upon, it could be dangerous, so go onto google maps or buy yourself a map and locate the pinpoint position of where you will be researching.

My number three rule is you can never pack too much.

When going out into the forest it is always good to pack a first aid kit, because you never know what might happen.
It might also be a good idea to pack a decent camera, just in case you come upon any potential evidence. Something that may be very useful also, is always carry a mobile device, so when you are out in the field you can contact people in case you need help.

Those are the absolute basic things you need to go out into the field. It's also a good idea if you pack some water and some food to keep your energy up, you never know if you might be out longer than you think.

Once you have completed your main tasks and you have a certain date you are going out, then you can go out and have fun researching, but never ever let your guard down, because it's when you let your guard down, that's when bad things happen.

When you are out in the forest safety is the key thing. You must always go with your phone and a first aid kit. It is also wise to download and app such as What3Words. If you get into trouble, and you need help, go into the app and it should give you three words which you can give to emergency services or someone else with the app and they can identify where you are.

It is always good to know what terrain you are going into because you then know what to expect and what you need to do in case of a slip, trip or fall.

You should always go prepared; you never know when something might happen. It is always key to know where you are going and always tell someone what time you are leaving and what time you are getting back. It may save your life.

My last piece of advice for new beginners when researching, is keep an open mind about everything you find and don't become attached to your research. When you have an open mind, you will find a lot more and you will experience a lot more.

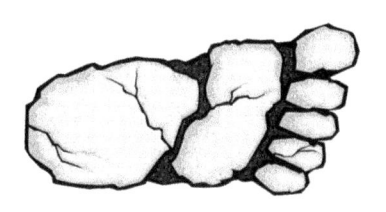

Chapter Nine
British Bigfoot

The British Bigfoot is just the same as every other type of Bigfoot in the world. The Bigfoot community has discredited the United Kingdom for cryptids such as Bigfoot. I can safely say that there is something not right going on here in the UK.

From findings and discoveries, I can say that some sort of creature is here that is yet to be discovered.

My theory about the British Bigfoot is based on the eDNA analysis that we did.

On our eDNA analysis there was a hit for old world monkey. Now, old-world monkeys lived here in the UK a while back and I thought about the possibility of them living on, and becoming what we know as today's British Bigfoot. As a researcher, you have to cover all the possible answers it could be, and unfortunately a lot of those answers are not answers you want, but do not give up, even if you are not finding anything.

In every state in America, they have a name for their local Bigfoot. Here in the United Kingdom, we don't have a specific name or anything that is native to this area. The reason is because there are not enough people out there looking for this myth, this legend, and if we had enough, I believe we would have the proof that America has.

I believe here in the UK we are dangerously close to what we are looking for, judged on the DNA and other evidence we have found.

One of the specific things I tell my team is "never base your research on other continents findings." The reason I say that, is because every country

My Bigfoot Life

has either similar things or completely different things, that have relations with Bigfoot. When we venture out into the forest, we base our research on what we have been told here in the UK, not in America or around the world.

It's OK to take some basic knowledge from around the globe, but to keep in mind that it could be completely different here in the UK.

The UK Bigfoot is a unknown creature that we don't know much about yet. Unlike in America, where they can make guesses judged on real evidence they have found. There is not much real evidence that hasn't been debunked yet here in the UK, which makes any findings even more important for the UK Bigfoot community.

Mythical Legends/Little Feet in the news

Explorer Daniel Barnett, 14, is on a mission to find Bigfoot

Daniel Barnett has struggled with autism since he was six years old, but his dad said his passion for the outdoors and Bigfoot had helped him a lot

Meet the UK's youngest explorer, who - at just 14 years old - is on a mission to find Bigfoot. Daniel Barnett has even started a podcast where he interviews famous scientists from across the globe in his search for the mythical creature.

He became fascinated with the subject after watching a documentary with his grandfather Adrian Roberts, 72. The plucky teen began to research the topic in depth, before setting up his podcast 'Mythical Legends' - where he interviews renowned experts and scientists. Daniel has autism and struggles with severe social anxiety, but his dad Craig, 44, who works in schools, claims that since starting his podcast he's become a "different person". The young explorer has set up cameras in Window Forest, Bridgwater, in the hope of tracking down the infamous Sasquatch and has even sent environmental DNA samples of a footprint he found to a lab in Portugal. Daniel hopes to encourage other young people to join him in his hunt for Bigfoot, who is said to be based in the US, by setting up an activities event with guest speakers and experts in the field.

Read the article here: https://www.walesonline.co.uk/news/real-life/explorer-daniel-barnett-14-mission-27913663

Daniel's on a mission to find Bigfoot

MEET the UK's youngest explorer who at just 14 years old is on a mission to find Bigfoot.

Daniel Barnett has even started a podcast where he interviews famous scientists from across the globe in his search for the mythical creature.

He became fascinated with the subject after watching a documentary with his grandfather Adrian Roberts, 72.

The plucky teen began to research the topic in depth, before setting up his podcast Mythical Legends – where he interviews renowned experts and scientists.

Daniel has autism and struggles with severe social anxiety, but his dad Craig

Barnett, 44, who works in schools, claims that since starting his podcast he's become a "different person".

The young explorer has set up cameras in Windown Forest, Bridgwater, in the hopes of tracking down the infamous Sasquatch and has even sent environmental DNA samples of a footprint he found to a lab in Portugal. Daniel hopes to encourage other young people to join him in his hunt for Bigfoot, who is said to be based in the US, by setting up an activities event with guest speakers and experts in the field.

Read the article here: https://www.pressreader.com/uk/western-daily-press/20231016/281560885457180

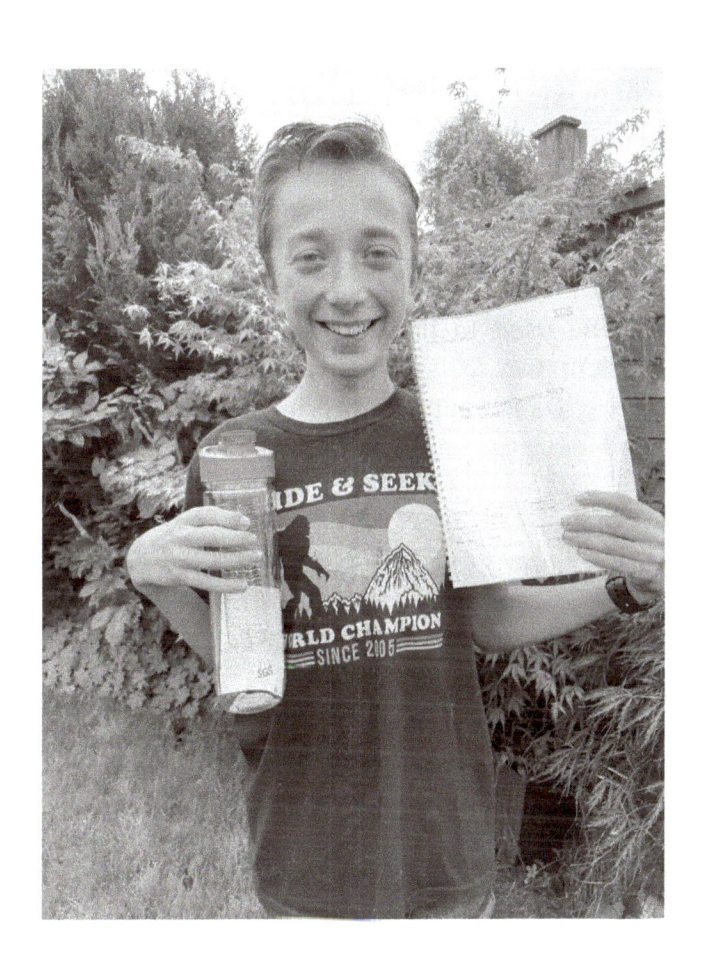

SGS Supports Young Daniel Barnett in Bigfoot Investigation

Meet Daniel Barnett: A 14-year-old who is quickly becoming one of the most famous Bigfoot hunters in the UK (Dailystar, Nottingham Post, Wales Online). We found out about him after he recently reached out to SGS with a very unusual request! Daniel runs a podcast in which he discusses mythical creatures, like Bigfoot. He also sets out camera traps in the local forest and goes on expeditions in search of large tracks. Daniel emailed us and wrote, "I collected soil samples from the tracks, and I would love if you could do DNA testing on them."

E-DNA Analysis SGS offers DNA testing for various sectors, but the most appropriate technology here involved Environmental DNA. E-DNA is the genetic material that organisms shed into the environment. It can be used to identify species in various environments and is ideal for finding elusive and rare animals.

"I was struck by the fact that this 14-year-old had made the link that e-DNA could maybe also be used to find the most elusive creature of them all: Bigfoot," says Willem van Strien, Innovation manager at the SGS Global Biosciences Center. "Daniel appeared to be so enthusiastic about the topic and really showed he was embracing science on his quest to get to the bottom of this."

Read the article here: https://www.sgs.com/en-gb/news/2023/10/sgs-supports-young-daniel-barnett-in-bigfoot-investigation

Teen hunting for British Bigfoot gets shock 'ape' DNA results after finding footprint

Teenager Daniel Barnett sent off DNA samples from a mysterious footprint he found while searching for Bigfoot in Britain - and the results suggest he might be onto something

A teenage Bigfoot hunter has received some intriguing results after finding a mystery footprint and sending DNA off for analysis.

Daniel Barnett, 14, has started a podcast 'Mythical Legends' - where he interviews famous scientists from across the globe in his search for the mythical creature. Daniel has been fascinated by Bigfoot since he and his grandfather watched a documentary.

The Daily Star previously reported that he had set up cameras in Windown Forest, Bridgwater, and even sent environmental DNA samples from a

My Bigfoot Life

footprint he found to a lab.

Read the full article here: https://www.dailystar.co.uk/news/weird-news/teen-hunting-british-bigfoot-gets-31235650?utm_source=mynewsassistant.com&utm_medium=referral&utm_campaign=embedded_search_item_desktop

Teen explorer to hold Bigfoot convention in HIghbridge this Saturday

A teenage explorer who is on the trail of Britain's Bigfoot is set to hold a special event in Highbridge this Saturday (November 18th).

Daniel Barnett, 14, who is autistic, recently started a podcast where he

My Bigfoot Life

interviews famous scientists from across the globe in his search for the mythical creature.

He is organising a 'Little Feet' convention at the YMCA Purple Spoon cafe in Highbridge from 10am until 2pm on Saturday when speakers from across Europe will be gathering. Father Craig Barnett says: "He's organised it all on his own, with guest speakers from all over Europe, book signings, fun activities for children and he will be raising money for the World Wildlife Fund."

"His aim is to encourage everyone to get out side and enjoy nature."

Some people are flying in from Portugal to attend.

Read the full article here: https://www.burnham-on-sea.com/news/teen-explorer-to-hold-bigfoot-convention-in-highbridge-this-saturday/

SGS
@SGS_SA

🦶 Meet Daniel Barnett, a 14-year-old #bigfoot hunter who collected soil samples from mysterious tracks in his local forest and reached out to us, hoping our E-DNA #analysis service could shed light on the matter. bit.ly/3tJTncU

Paranormal UK Radio Network @PAUKRadio · Oct 11
📢 New Podcast! "Paranormal UK Radio Show - Mythical Legends with **Daniel Barnett**" on @Spreaker #allen #barnett #bigfoot #cryptids #daniel #dogman #irene #johnson #legends #mark #mythical #paranormal #radio #uk

spreaker.com
Paranormal UK Radio Show - Mythical Legends with
Irene and Mark talk with 14 year old cryptid researcher and podcaster Daniel Barnett from the ...

> **Cambridge Skeptics** @cambskeptics · Nov 1
>
> Teenager **Daniel Barnett** sent off DNA samples from a mysterious footprint he found while searching for **Bigfoot** in Britain, and the results suggest he might be onto something, report @jaymehudspith and @EthanBlackshaw for @dailystar.

SO EP:389 Bigfoot Discovered In The UK?
Sasquatch Odyssey

My guest tonight is Daniel Barnett At just 14 years old, Daniel is the youngest explorer in the UK, and he's on a quest to find Bigfoot. He has even launched a podcast where he converses with eminent scientists worldwide in his pursuit of the legendary creature. His interest in the subject was...
read more

🕒 19 days ago #barnett, #bigfoot, #cryptid, #cryptozoologist, #daniel, #dogman, #encounter, #kingdom, #podcast, #sasquatch, #story, #united, #weird, #youngest

Here the podcast here: https://www.spreaker.com/user/14234225/so-ep-389-bigfoot-discovered-in-the-uk

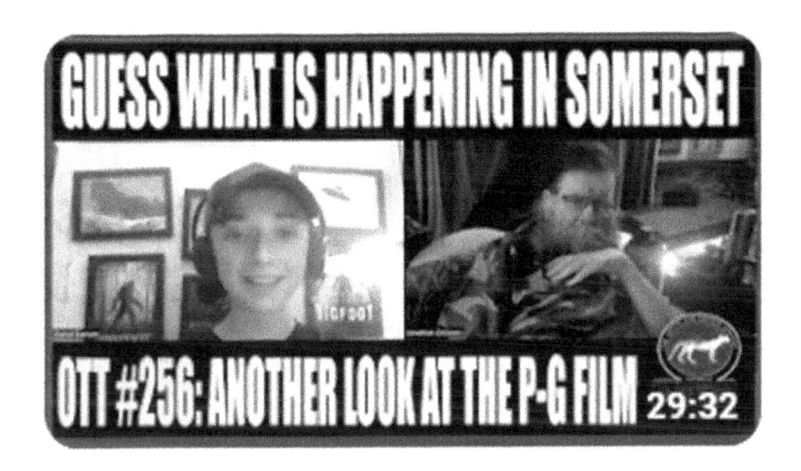

OTT #255: Young Somerset Crypto research, and another look at the...

Watch Daniel on CFZ's On The Track here: https://www.youtube.com/watch?v=xeA4PRLA2Ww

OTT #256: Daniel Barnett and Monster Hunting in Somerset (Par...

Watch Daniel on CFZ's On The Track here: https://www.youtube.com/watch?v=vuApcVdrZaw

My Bigfoot Life

OTT #261: UK BIGFOOT - What Daniel did next

Watch Daniel on CFZ's On The Track here: https://www.youtube.com/watch?v=7Yd1nSEduTQ&t=26s

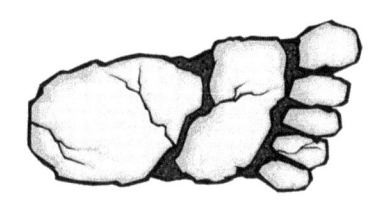

Chapter Ten
What to do if I come across evidence?

When going out researching, it's always good to know who to contact if you come across some possible evidence, that you want some of us to have a look at.

There is a group of good people who will help with your evidence.

If you ever came across some evidence, you think is worth a proper investigation, this advice will help a lot. Some of the greatest advice I was given, was never give out the location of your evidence, only to people you trust and people that are there to help you. Unfortunately, there are some people out there who want your evidence and will take it as their own.

So, before you let anyone know, make sure you know who the person is and what he does in the Bigfoot community.

Some suggestions for contacts are the UKBRT[10], this is my team, and it stands for UK Bigfoot Research team. We are working with researchers from up and down the UK. Chris Allsford is Co-Founder of the UKBRT, and he is the man to contact if you wanted to report a sighting to us. You can also contact the CFZ[11], they would really take in your story and help you with your evidence.

There are a lot of good people out there you just have to go and find them. Always make sure they are the right people.

You are more than welcome to contact me, I may be busy, but you are more than welcome to reach out.

---

[10] https://www.facebook.com/profile.php?id=61553030703803
[11] https://cfz.org.uk/

If you wanted to handle the evidence yourself, then I would suggest taking photos of the evidence and maybe collect a sample of eDNA for testing and remember you spot you found it in.

Casting is another piece of knowledge which is brilliant to know because if you come across a footprint you can cast the print, which means you have a three dimensional model of the footprint which means that later down the line you can check for dermal ridges and toe separation on the cast itself.

When you are casting a print you need to make sure you use a big enough bucket and you need to ensure you have a medium size bottle of water. Sou ensure you have those items and your bag of Plaster of Paris.

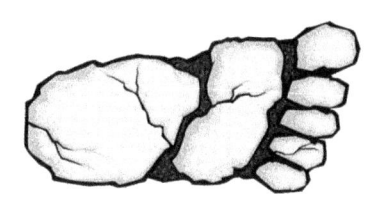

Chapter Eleven
The Future

The future for my work is a whole new universe. My plans for the bigfoot community are huge, when people go out into town and they say the word Bigfoot, they talk to you about what they have heard or seen, not say you are crazy and delusional. As a team together we have big plans to help put Bigfoot on the UK map, starting with a yearly book!!

We are planning to write a book every year to explain the work that went on that year and the interesting revelations,

Our books will be based on what is happening here in the UK and the surrounding area. We hope to have some very interesting things inside and if you are coming to the end of this one, then you are going to love the next one.

The future for the Bigfoot Conventions is absolutely certain. We are going to be continuing to bring people, to create a bigger community that can speak their thoughts about what they think cryptids are and their feelings towards them. Our future for the podcast is to continue as it's going and to see where it leads us. We have already had some big names pop up on our podcast, like Dr Jeffrey Meldrum, an absolute legend.

So exciting things are coming up.

***

Without Jonathan Downes this book wouldn't be here. The UKBRT wouldn't be the UKBRT without him. He brings a sense of comedy and joy to the team and he is a tremendous asset. We are very lucky to have someone like him. Thank you to you Jon; you are an excellent man.

My Bigfoot Life

THE WORLD'S WEIRDEST PUBLISHING COMPANY

HOW TO START A PUBLISHING EMPIRE

Unlike most mainstream publishers, we have a non-commercial remit, and our mission statement claims that "we publish books because they deserve to be published, not because we think that we can make money out of them". Our motto is the Latin Tag *Pro bona causa facimus* (we do it for good reason), a slogan taken from a children's book *The Case of the Silver Egg* by the late Desmond Skirrow.

WIKIPEDIA: "The first book published was in 1988. *Take this Brother may it Serve you Well* was a guide to Beatles bootlegs by Jonathan Downes. It sold quite well, but was hampered by very poor production values, being photocopied, and held together by a plastic clip binder.

In 1988 A5 clip binders were hard to get hold of, so the publishers took A4 binders and cut them in half with a hacksaw. It now reaches surprisingly high prices second hand.

The production quality improved slightly over the years, and after 1999 all the books produced were ringbound with laminated colour covers. In 2004, however, they signed an agreement with Lightning Source, and all books are now produced perfect bound, with full colour covers."

Until 2010 all our books, the majority of which are/were on the subject of mystery animals and allied disciplines, were published by `CFZ Press`, the publishing arm of the Centre for Fortean Zoology (CFZ), and we urged our readers and followers to draw a discreet veil over the books that we published that were completely off topic to the CFZ.

However, in 2010 we decided that enough was enough and launched a second imprint, `Fortean Words` which aims to cover a wide range of non animal-related esoteric subjects. Other imprints will be launched as and when we feel like it, however the basic ethos of the company remains the same: Our job is to publish books and magazines that we feel are worth publishing, whether or not they are going to sell. Money is, after all - as my dear old Mama once told me - a rather vulgar subject, and she would be rolling in her grave if she thought that her eldest son was somehow in `trade`.

Luckily, so far our tastes have turned out not to be that rarified after all, and we have sold far more books than anyone ever thought that we would, so there is a moral in there somewhere…

Jon Downes,
Woolsery, North Devon
July 2010

CFZ PRESS

CFZ Press is our flagship imprint, featuring a wide range of intelligently written and lavishly illustrated books on cryptozoology and the quirkier aspects of Natural History.

CFZ Classics is a new venture for us. There are many seminal works that are either unavailable today, or not available with the production values which we would like to see. So, following the old adage that if you want to get something done do it yourself, this is exactly what we have done.

Desiderius Erasmus Roterodamus (b. October 18th 1466, d. July 2nd 1536) said: "When I have a little money, I buy books; and if I have any left, I buy food and clothes," and we are much the same. Only, we are in the lucky position of being able to share our books with the wider world. CFZ Classics is a conduit through which we cannot just re-issue titles which we feel still have much to offer the cryptozoological and Fortean research communities of the 21st Century, but we are adding footnotes, supplementary essays, and other material where we deem it appropriate.

http://www.cfzpublishing.co.uk/

Fortean Words is a new venture for us. The F in CFZ stands for "Fortean", after the pioneering researcher into anomalous phenomena, Charles Fort. Our Fortean Words imprint covers a whole spectrum of arcane subjects from UFOs and the paranormal to folklore and urban legends. Our authors include such Fortean luminaries as Nick Redfern, Andy Roberts, and Paul Screeton. . New authors tackling new subjects will always be encouraged, and we hope that our books will continue to be as ground-breaking and popular as ever.

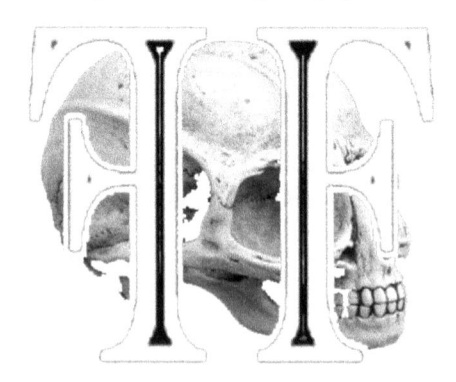

Just before Christmas 2011, we launched our third imprint, this time dedicated to - let's see if you guessed it from the title - fictional books with a Fortean or cryptozoological theme. We have published a few fictional books in the past, but now think that because of our rising reputation as publishers of quality Forteana, that a dedicated fiction imprint was the order of the day.

http://www.cfzpublishing.co.uk/

Milton Keynes UK
Ingram Content Group UK Ltd.
UKHW022027100524
442551UK00006B/358